作者简介

张立，北京大学大数据科学研究中心助理研究员，美国艾奥瓦大学博士。研究方向为面向图像数据的人工智能算法、多模态医学影像分析、多源异构医疗数据处理等，致力于推进智能技术与深度学习算法在医疗领域的应用。在 *JAMA Ophthalmology*、*IEEE Transactions on Medical Imaging* 等 SCI 期刊，以及 MICCAI、ICCV、AAAI 等国际顶级会议上发表专业论文 30 余篇。主持和作为课题骨干参与国家自然科学基金重大项目与青年项目、北京市自然科学基金重大项目、科技部重点研发计划等重要科研项目。

About the Author

Li Zhang is an assistant researcher at the Center for Data Science of Peking University. He has earned a Ph. D. from the University of Iowa. His research areas include artificial intelligence algorithms for image data, multi-modal medical image analysis, and multi-source heterogeneous medical data processing. Zhang is committed to promoting the application of intelligent technology and deep learning algorithms in the medical field. He has more than 30 publications in the world-class SCI journals (such as *JAMA Ophthalmology* and *IEEE Transactions on Medical Imaging*) and top international conferences such as MICCAI, ICCV, and AAAI. In addition, Zhang has served as the investigator/co-investigator of a couple of important scientific research projects including the Major Program and the Youth Program supported by National Natural Science Foundation of China, the Major Program supported by Beijing Municipal Natural Science Foundation, and the Key Research and Development Plan of the Ministry of Science and Technology of the People's Republic of China.

Handbook for the Construction of Makeshift Hospitals

方舱医院建设工作手册

（中英双语版）

张　立　主编

科学技术文献出版社
SCIENTIFIC AND TECHNICAL DOCUMENTATION PRESS
·北京·

图书在版编目(CIP)数据

方舱医院建设工作手册：汉、英 / 张立主编. —北京：科学技术文献出版社，2021. 3

ISBN 978-7-5189-7753-6

Ⅰ. ①方… Ⅱ. ①张… Ⅲ. ①传染病医院—建筑工程—手册—汉、英 Ⅳ. ① TU246.1-62

中国版本图书馆 CIP 数据核字（2021）第 060945 号

方舱医院建设工作手册（中英双语版）

策划编辑：周国臻　责任编辑：宋红梅　责任校对：王瑞瑞　责任出版：张志平

出 版 者　科学技术文献出版社
地　　址　北京市复兴路15号　邮编　100038
编 务 部　（010）58882938，58882087（传真）
发 行 部　（010）58882868，58882870（传真）
邮 购 部　（010）58882873
官方网址　www.stdp.com.cn
发 行 者　科学技术文献出版社发行　全国各地新华书店经销
印 刷 者　北京虎彩文化传播有限公司
版　　次　2021 年 3 月第 1 版　2021 年 3 月第 1 次印刷
开　　本　880 × 1230　1/32
字　　数　64千
印　　张　4　彩插 2 面
书　　号　ISBN 978-7-5189-7753-6
定　　价　38.00元

编　委　会

Editorial Board

武汉方舱医院休舱仪式
2020年2月5日-3月10日
南1
科
全
国
胜
湖
北
武
汉
胜

序言（一）

毛宗福

2019年12月，湖北省武汉市暴发了新型冠状病毒肺炎（简称“新冠肺炎”）疫情，疫情逐步肆虐全国，引发全球高度关注。在习近平总书记的亲自指挥、亲自部署下，全国各界紧张有序、有条不紊地进行了抗疫斗争。至2020年4月8日，武汉全面解封，逐步恢复正常社会生活，标志着我国抗疫工作取得了伟大胜利。

方舱医院的创设，是此次抗击新冠肺炎疫情的关键举措，在“防”与“治”两个方面发挥了不可替代的重要作用。当疫情形势最为严峻时，病患就医数量呈“井喷式”增长，大量患者在社区和社会流动；与此同时，医疗资源极度紧张，床位不能满足“应收尽收”的要求，面临延误治疗时机、造成疫情扩散的双重压力。关键时刻，党中央果断做出建设方舱医院的决策部署，武汉市将一批体育场馆、会展中心等改造为方舱医院，在短时间内迅速解决了床位不足的问题，大大加快了患者的收治力度，有效降低了轻症向重症的转化率。

从2020年2月3日首家方舱医院建设开始，到2月27日出现“床等人”现象，历时仅20多天，就彻底改变了“一床难求”的局面。据统计，各家方舱医院在运行30多天里共计收治新冠肺炎轻症患者12 000多人，成为名副其实的“生命之舱”。

采用大规模的方舱医院来防控疫情，是我国公共卫生防控与医疗的一个重大创举，包含着鲜明的中国特色和中国智慧；它背后的运行机制也是创造，归根结底，是党中央集中统一领导下制度优势支撑起来的创举。党中央一声令下，武汉即刻启动建设，全国一盘棋、四面八方驰援，一座座方舱医院从无到有、从设想成为现实。第一批有4000张床位的3家方舱医院仅用29小时就建设完成。中央赴湖北指导组调动22支国家紧急医学救援队及车载方舱、3辆国家移动核酸检测实验车星夜驰援武汉，76支医疗队、8000多名医务人员在几天内陆续进入方舱医院，边建设、边接收、边治疗，迅速开展救治工作。从场馆施工改造，到快速收治患者入院，到建立一系列流程、岗位、文书等制度，明确医护人员岗位职责、各就各位，到完善诊疗流程，建立快速、畅通的重症转诊机制等，方舱医院逐步实现精细化管理、平稳高效运行。方舱医院从无到有、有效运转、不辱使命的30多天，充分表明党的领导是中国特色社会主义制度的最大优势，党中央一声号令，全党同志闻令而动、全国人民上下同心，迸发出攻坚克难、创造性解决问题的强大力量，最大限度地调集有效资源，在最短时间内找到了克服危机的最佳办法，并有效地得以实施，完成了既定目标。

疫情防控要求相关各方都能做到守土有责、守土担责、守土尽责。面对来势汹汹的新冠肺炎疫情，为了进一步认真落实党中央和国务院关于新型冠状病毒肺炎疫情防控工作的决策部署，切实做好在未来疫情可能长期存在或有所反复情况下的综合应对工作，武汉市新冠肺炎疫情防控指挥部组织相关专家，结合疫情期间武汉地区的防控经验，参考国内外相关文献证据，在充分征求全国相关专家意见的基础上，编写了《方舱医院建设工作手册（中英双语版）》。本书内容丰富、科学全面，从方舱建设的技术指导原则、管理制度设计到相关软硬件、设施和设备建设，都做了充分阐述和介绍，不仅全面总结了武汉抗疫战斗中方舱医院建设的经验与不足，也为全国乃至世界各地应对未来可能反复发生的新冠肺炎疫情甚至各类重大公共卫生危机时组织方舱医院建设提供了一般性管理思路和具体解决技术方案，具有重大的社会医学与公共卫生事业价值。

（作者系武汉大学全球健康研究中心主任，教授、博士生导师）

Preface(I)

Zongfu Mao

In December 2019, a new type of coronavirus pneumonia—COVID-19 broke out in Wuhan, Hubei Province. The epidemic gradually raged across the country and aroused global attention. Under the command and deployment of President Xi Jinping, the whole nation fought against the epidemic in an orderly manner. As of April 8, 2020, the citywide lockdown of Wuhan ended and the regular civil life resumed, symbolizing a periodical victory of China's anti-epidemic work.

The construction of makeshift hospitals was a pivotal measure in fighting against the COVID-19 epidemic due to its irreplaceable function in both "prevention" and "treatment". The number of patients seeking medical treatment went into the blowout of a sharp increase along with the expansion of epidemic. Patients had to transit between resident communities and public places where hospitals are generally located. In addition, medical resources in Wuhan were extremely scarce to provide sufficient beds to meet the needs. The local government fell into the double pressure of missing best treatment opportunity and risking extensive infections. In this context, the Communist Party of China (CPC) Central Committee made a timely and decisive order to build makeshift hospitals in Wuhan. Subsequently, a number of stadiums and convention centers were turned into makeshift hospitals. This measure effectively resolved the contemporary insufficiency of beds, which accelerated the admission for treatment and reduced the conversion rates of patients from mild to severe conditions.

Specifically, on February 3, 2020, the construction of the first makeshift hospital began in Wuhan. By February 27, 2020, the number of hospital beds greatly increased and the situation of "Patients Waiting for One Bed" (i. e., the shortage of hospital beds) in Wuhan had completely changed to "Beds Awaiting for Patients" (i. e. , the enough supply of hospital beds) in merely over 20 days upon the construction of makeshift hospitals. After a month of the operating, over 12,000 patients with mild COVID-19 were registered and treated in the makeshift hospitals, which justifies the reason why they are deserved to be named as "the Cabins of Life" in China.

Therefore, adopting large-scale makeshift hospitals to prevent and control epidemics is a vital, pioneering action to ensure accountable public health response and medical treatment provision in China. First, it reveals the exceptional advantages and wisdom of China embedded in its institutional mechanism—rapid and efficient resource mobilization under the centralized and unified leadership of the CPC Central Committee. Under the order from the CPC Central Committee, Wuhan immediately launched the construction of makeshift hospitals with the support from the entire country. Every corner in China was in solidary, disciplined and cohesive action: one makeshift hospital after another was constructed with "unbelievable speed"—the first 3 makeshift hospitals with 4,000 beds only took 29 hours; 22 national emergency medical rescue teams and mobile cabins and 3 national mobile nucleic acid testing experimental vehicles straightly head off into Wuhan to provide support; 76 medical teams with more than 8,000 medical personnel arrived Wuhan and entered the makeshift hospitals within only few days. As a result, the construction process of makeshift hospitals was managed to synchronize with the registration and treatment of patients without delay. Second, the establishment of a series of standardized management regulations on the division of job responsibility, the process of diagnosis and treatment, and the mechanism of primary healthcare referrals, etc. , ensured the stable and cohesive operation of the makeshift hospitals. During these 30 days, we witnessed how these makeshift hospitals were transformed from scratch construction to effective function to achieve their mission. Such leadership of the CPC is the greatest advantage of the Socialism with Chinese Characteristics. Under the command of the CPC Central Committee, full comrades of the party and people of the entire nation came together with one direction, bursting out the power of solidarity and creativity to overcome this overwhelming hard time.

Under the guidance of CPC Central Committee on the prevention and control of COVID-19, the Wuhan headquarters compiled the *Handbook for the Construction of Makeshift Hospitals* as an evidence-based reference for effective response to future coronavirus pandemics. Grounded by the learnings and successful experiences of prevention and control in Wuhan, the compilation of this workbook has comprehensively integrated with relevant theories and practices from domestic and foreign literature, as well as the insights from topnotch public health experts. It offers a detailed technical guideline and principles for China and even for the world, shedding a light on the process of the design and construction of makeshift hospitals, ancillary software and hardware facilities and equipment, as well as corresponding management system. The workbook has the potential to help the city, the nation and the world to better respond to the coronavirus pandemic and various public health crises in the future. Wuhan's practices and experiences in the battle against COVID-19 would be exclusive and significant references for the post-pandemic reconstruction of the social medicine and public health system.

(The author is the director of Global Health Institute of Wuhan University, professor and doctoral supervisor.)

序言（二）

马占军

新冠肺炎（COVID-19）疫情是百年来全球发生的最严重的传染病大流行，是新中国成立以来我国遭遇的传播速度最快、感染范围最广、防控难度最大的重大突发公共卫生事件。随着对新冠病毒的逐步了解，人们发现，感染者中轻度患者的比例约占 80%。为隔离控制传染源，武汉在世界上首次将方舱医院用于急性传染病的治疗和隔离，是一项重大举措和创新。2020 年 2 月 28 日，在武汉举行的新冠肺炎疫情防控工作新闻发布会上，国家卫生健康委主任马晓伟将方舱医院的建立称为在医疗资源紧张、疫情扩散压力下“极其正确、果断的决定”。

从 2020 年 2 月 3 日开始，武汉先期 10 余天建成 16 家、陆续共建设 32 家方舱医院，合计建成床位 32 868 张。到 3 月 10 日最后一家方舱医院休舱，累计收治患者1.2万余人，治愈出院8000余人，转院3500余人，做到了“零感染、零死亡、零回头”。方舱医院实现了“应早尽早”收治确诊轻症患者的目标，有效防止了轻症患者因得不到治疗而转变成重症，隔离了传染源，遏制了疫情扩散，成功减轻了定点医院的床位压力。

武汉建设方舱医院的主要推动者，中国工程院副院长、中国医学科学院院长王辰院士团队，和德国海德堡大学全球公共卫生研究所等团队，共同在全球顶级医学期刊《柳叶刀》（*The Lancet*）线上发表了一篇聚焦中国建设和启用方舱医院（makeshift hospitals）的文章。他们在文章中总结，方舱医院有3个特点：快速建设，大规模，建设和运行成本较低；方舱医院实现了5个基本功能：隔离，分诊，提供基本医疗服务，频繁监测和迅速转诊，提供基本的生活和社会交往。笔者认为，方舱医院是中国成功控制COVID-19疫情的主要原因。中国在应对COVID-19疫情时启用方舱医院的经验表明：未来一旦突发公共卫生事件，方舱医院可以被强有力地采用。尤其适用于其他的流行病，也可以用于其他大规模或快速增长的涉及疾病或损伤的事件，如大规模中毒或自然灾害。

尽管方舱医院所收治的患者主要为轻型或普通型患者，但是病患数量巨大，除了对新冠肺炎的一般治疗，同时也要兼顾其他伴随慢性病的治疗，并且既要注重轻症治疗也要考虑突然转重患者的急救，所以方舱医院对医用设备、医用药品及医用耗材的需求迫切而且特殊。鉴于此，武汉市新冠肺炎疫情防控指挥部应急及物资保障组通过不断摸索并多次修订，编写了《方舱医院建设工作手册》，为中国乃至全球应对突发大型公共卫生事件提供了“武汉经验”。

作为防疫战略的一部分，方舱医院需要多种支持和解决方案，其中医药保障是重要一环。在武汉方舱医院建设和运营过程中，医药企业发挥了不可或缺的作用。整个方舱医院运营期间，以中央在汉大型医药企业国药控股湖北有限公司为代表的湖北、武汉市场主要医药企业，发扬“并肩抗疫、药企担当”的精神，按照中央赴湖北指导组的部署和省、市疫情防控指挥部工作要求，主动承担社会责任，全力做好抢险救灾药品及防疫物资的储备、调拨和供应保障工作任务。医药人不惧风险、冲锋在前，克服人员紧张、物资不足、风险性大等困难，全力保障了包括方舱医院在内的抗疫一线的应急医疗物资需求，为武汉保卫战、湖北保卫战贡献了巨大力量。

目前，抗击新冠肺炎疫情斗争已取得重大战略成果。为深入学习贯彻习近平总书记在全国抗击新冠肺炎疫情表彰大会上的重要讲话，弘扬伟大抗疫精神，总结方舱医院建设运营经验，作为方舱医院建设运营的参与者、见证者，我们将《方舱医院建设工作手册》修改补充后，正式以中英文双语的形式出版发行，期望对当下和今后世界范围内的疫情防控有所借鉴。

（作者系湖北省人大代表、湖北省人民政府参事、湖北省医药行业协会名誉会长，华中科技大学兼职教授、中南财经政法大学客座教授，国药控股湖北有限公司董事长）

Preface(II)

Zhanjun Ma

In the past hundred years, COVID-19 is the most serious infectious disease pandemic worldwide. It is a major public health emergency with the fastest transmission speed, the widest infection range and the greatest difficulty in prevention and control since the founding of the People's Republic of China. With the gradual understanding of the novel coronavirus, it was found that the proportion of infected patients with mild symptoms accounted for approximately 80% . For the first time in the world, Wuhan applied the makeshift hospitals in the treatment and isolation of acute infectious diseases, which is a significant measure and innovation. At the news conference on COVID-19 epidemic held in Wuhan on February 28, 2020, Ma Xiaowei, the director of the National Health Commission, called the establishment of the makeshift hospital "an extremely correct and decisive decision" under the great pressure of medical resources shortage and epidemic spread.

Since February 3, 2020, 16 makeshift hospitals in the first phase have been built in more than ten days. A total of 32 makeshift hospitals were built one after another, with a total of 32,868 beds. The last makeshift hospital in Wuhan closed on March 10, 2020. During the opening period of the makeshift hospitals in Wuhan, more than 12,000 patients were treated, over 8,000 patients were cured and discharged, more than 3,500 patients were transferred to another hospital, and the achievement of "no infection, no death, and no return" had been made. The construction of makeshift hospitals achieved the goal of diagnosing and treating mild patients as early as possible, which effectively prevented the disease progression of mild to severe due to the lack of treatment, isolated the infection source, curbed the epidemic spread, and successfully relieved the bed pressure of designated hospitals.

Academician Wang Chen, vice-president of the Chinese

Academy of Engineering and president of the Chinese Academy of Medical Sciences, is the main promoter for the construction of makeshift hospitals in Wuhan. His team, along with a research group from the Global Institute of Public Health, Heidelberg University, et al. , published a paper in *The Lancet* which focused on the construction and use of Chinese makeshift hospitals. They concluded that makeshift hospitals had three characteristics: rapid construction, large scale, and low costs of construction and operation; achieved five basic functions: isolation, triage, provision of basic medical services, frequent monitoring and rapid referral, and provision of basic life and social interaction. The authors believe that the construction of makeshift hospitals is the main reason for the effective control of COVID-19 epidemic in China. China's experience of using makeshift hospitals in COVID-19 epidemic shows that makeshift hospitals can be forcefully adopted to cope with public health emergency in the future. The makeshift hospital is especially applicable to epidemics, as well as other large-scale or rapidly growing events involving disease or injury, such as mass poisonings or natural disasters.

Although the patients admitted to makeshift hospitals were mainly mild or general patients, the total number of patients was huge. In addition to the general treatment of patients with COVID-19, the treatment of other concomitant chronic diseases should be considered. Also, it should not only pay attention to the treatment of mild symptoms, but also consider the first aid of patients who suddenly turn to severe symptoms. Therefore, makeshift hospitals had urgent and special demands for medical equipment, medical drugs and medical consumables. Under the circumstance, the Emergency Response and Material Support group of Wuhan COVID-19 Prevention and Control Headquarters developed the *Handbook for the Construction of Makeshift Hospitals* by constant exploration and repeated revision, which provides "Wuhan Experience" in dealing with great public health emergencies for China and even the whole world.

The makeshift hospital is a part of the overall strategy for COVID-19 prevention and control and requires multiple supports and solutions, among which medical and pharmaceutical support matters. Pharmaceutical enterprises played indispensable roles in the process of constructing and operating makeshift hospitals in Wuhan. During the entire operation period of makeshift hospitals, the major pharmaceutical enterprises in Hubei and Wuhan markets, represented by the Sinopharm Group Hubei Co. Ltd. , actively participated in the anti-epidemic work. In accordance with the deployment of the CPC Central Committee, Hubei provincial and Wuhan COVID-19 prevention and control headquarters, pharmaceutical enterprises spared no effort to ensure the reserve, allocation and supply of emergency and relief drugs and epidemic prevention materials. The related personnel overcame multiple difficulties (personnel shortage, material shortage, high risk of infection, etc.), and took all efforts to ensure the needs of emergency medical materials on the front line of anti-epidemic including the makeshift hospitals, and made great contributions to the "Defensive Battle of Wuhan and Hubei".

At present, significant strategic achievements have been made in the COVID-19 anti-epidemic work. As the participants and witnesses of the construction and operation of the makeshift hospitals in Wuhan, we revised the *Handbook for the Construction of Makeshift Hospitals* and published in both Chinese and English. This workbook summarized the experience of constructing and operating makeshift hospitals in Wuhan, trying to provide references for the prevention and control of COVID-19 epidemic worldwide at the moment and in the future.

(The author is a deputy of Hubei Provincial People's Congress, counselor of Hubei Provincial People's Government, honorary president of Hubei Pharmaceutical Industry Association, adjunct professor of Huazhong University of Science and Technology, guest professor of Zhongnan University of Economics and Law, and chairman of China Pharmaceutical Control Hubei Co., Ltd.)

目　　录（中文）

Contents (English)

第一章　方舱医院建设预备工作

一、组织架构

（一）成立方舱医院应急及物资保障组

应急及物资保障组领导小组总体负责方舱医院医用物资管理、调配、采购和配送相关工作，对人员、防护用品、设备、药品和耗材统一计划与调配。具体包括测算医用物资需求、制定/审核医用物资采购方案、完成医用物资的供应与调配等。

（二）应急及物资保障组下设不同职能小组

应急及物资保障组领导小组针对医用物资管理，设立物资需求与分配组、物资采购与调拨组、财务统计组、管理审核组，各小组各司其职，为方舱医院运行提供保障。各成员在防控期间处于 24 小时运行状态，随时准备应对突发事件。各小组具体职能如下。

1. 物资需求与分配组

负责对接各家方舱医院的物资情况，包括防护用品、医用耗材、药品、医疗设备、消杀用品等，并统计具体的物资需求和运行后的物资缺口。根据需求与缺口的统计结果，制订分配计划，负责对接相应物资（包括捐赠物资）并进行分配，提供分配表。

2. 物资采购与调拨组

根据需求表负责采购物资并将其送达方舱医院。负责对送达的设备进行安装和对使用人员进行培训。同时，本组也负责物资的运输、物流等的统筹工作。

3. 财务统计组

负责对防护物资、医用耗材、药品、设备等重要物资的登记和信息管理工作。

4. 管理审核组

负责对财务、物资配置标准、物资质量控制标准进行审核。负责对采购物资和捐赠物资的质量进行把关，查验合法资质、效期等验收工作。严格执行国家医用设备、医用耗材和药品相关监管规定，控制各类物资的质量，发现失效或者不合格产品应停止使用，防止此类产品用于防疫前线。

全面部署，全力保障（国药控股湖北有限公司供稿）

二、方舱医院的选址与建设标准

方舱医院作为应对严重疫情迅速建立的大规模临时医疗场所，能够及时提供大容量、低成本的简易医疗床位，使大量确诊轻症患者能够及时得到有效治疗。建设方舱医院，可防止轻症患者得不到有效治疗转变成重症，既能减轻定点医院床位压力，又能有效隔离传染源，遏制疫情扩散。

（一）方舱医院的选址

方舱医院的选址应该满足以下条件：

①方舱医院原则上选择已经有水电交通、配套功能齐全的建筑，保证方舱内收治和隔离过程中的基本生活条件，同时也保证医用物资的运输和医护等工作人员的通勤方便。

②方舱医院应尽量为独立单体建筑，与周边居民区有充足的间距，避免医用垃圾对周边居民区产生污染。

③有独立的污水处理系统或能快速改造污水排水管。

④根据《临时改建方舱医院消防安全要求（试行)》，按照床位数量，方舱医院应至少具有两个以上的进出交通通道。

⑤建筑面积单层不低于 1 万平方米，在保证患者有相对独立的使用空间的前提下，尽可能安排床位。

以新型冠状病毒肺炎为例，武汉市前期建设的方舱医院主要选择予以改造的建筑，包括 10 家体育馆、4 家展览馆、2 所高等院校和 1 家工业园区。体育馆是方舱医院选址改造的首选，原因有二：首先，体育馆是独立单体建筑，改造方便，进出口多，便于达到院感控制要求；其次，体育馆的空间足够大，并且改造成本相对较低。

（二）方舱医院的建设标准

严格执行“三区两通道”的标准。污染区、半清洁区和清洁区严格分开，污染通道和清洁通道合理排布，如下图所示。除此以外，对于方舱医院的建设，应满足《关于调整方舱医院医用物资开舱配置标准的通知》中关于方舱医院开舱的配置标准，并予以匹配相应的消防安全要求，参考湖北省消防救援总队研究出台的《临时改建方舱医院消防安全要求（试行）》（见附录），在信息系统建设、防火、防水、故障解决上进行设计。

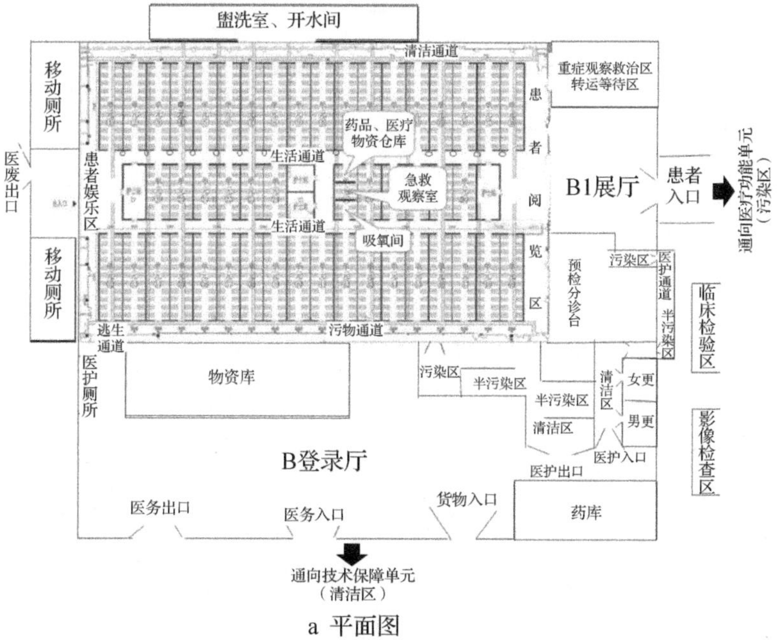

a 平面图

b 实物图

武汉国际博览中心方舱医院

第二章　人员与医用物资配置标准

一、方舱医院人员与医用物资配置标准总览

按照《湖北省方舱医院救治工作管理规范》相关要求，根据武汉市新型冠状病毒肺炎防控的实践经验，各级各类医院防控一线的医务人员需求，以及火神山、雷神山、前期方舱医院筹备的建设标准，并结合方舱医院收治轻症患者的实际情况，武汉市新冠肺炎疫情防控指挥部于2020年2月20日印发《关于调整方舱医院医用物资开舱配置标准的通知》，初步确定了方舱医院人员与医用物资配置标准。该标准以1000张床位为基础单位，规定500张床位以上的方舱医院医用设备、医用药品及医用耗材的配置标准参照1000张床位执行，500张床位以下的酌情减量配置，减量的物资数量以向上取整为原则，尽可能满足各家方舱医院的需求，并且做到合理调配、避免冗余。方舱医院医用物资配置标准总体如下。

（一）人员配置标准

方舱医院原则上每收治100名患者配备10名医生、

40名护士；每收治100名患者配备1名西药师，每收治500名患者配备1名中药师；每收治50名患者配备1名院感人员。每收治200名患者配备1名治安民警（或第三方保安）、每收治500名患者配备1名消防管理人员、每收治100名患者配备4～6名保洁人员、每10个厕所配备1名保洁人员。

（二）防护物资配置数量标准

对于以上方舱医院中人员，应符合每人次每日配备一套符合要求的紧急防护服、医用防护口罩（N95）和一次性护目镜或防护面罩，一次性手套随时更换。防护物资可按照每收治1000名患者每天约配备600套的标准，总体对符合要求的紧急防护服、医用防护口罩（N95）、一次性医用手套按需配备。

（三）物资配置品目

按1000张床为计算标准，防护物资包括一次连体性防护服（带鞋套）、医用防护口罩（N95）、无菌手套（外科手套）等11种；医用耗材包括呼吸管路、快速手消毒液、75%医用酒精等46种；医用药品包括吸入用布地奈德混悬液、盐酸莫西沙星片、盐酸阿比多尔片、中药方剂等97种按10日需求量计算；医用设备包括心电监护仪、除颤仪、呼吸机（无创）等33种。

应急及物资保障组负责方舱医院医用物资管理相关

工作，实行人员、防护用品、设备、药品和耗材统一计划与调配，在本章中将说明人员、防护用品、设备、药品和耗材的配置与需求测算依据。

向方舱医院提供充足医药保障（江苏恒瑞医药股份有限公司供稿）

二、人员配置

方舱医院作为应对严重疫情迅速建立的大规模临时医疗场所，能够及时提供大容量、低成本的简易医疗床位，为大规模的确诊轻症患者提供有效治疗和隔离场所。参考《医疗机构基本标准》中相关医院的人员配置标准和原则，针对方舱医院做出如下配置。

1. 医生

方舱内设多个病区，一个病区设置 50 张病床，相应配备5名医生。每天5名医生中的3名进入舱内污染区进行日常医疗服务（穿戴全套防护用品），一人一次8小时，三班轮转。另外2名医生一人充当责任医生，负责5人医疗小组的总体事务，每天早晨进舱查房一次，另一人作为机动人员随时准备替班。

2. 护士

每一个病区（50张床位）配备20名护士。每天20名护士中的18名分为6组进入污染区进行日常医疗服务（穿戴全套防护用品），每组3名护士，一次工作4小时，6个班次轮转。剩余2名护士在清洁区进行后勤保障工作。每个病区配备1名护士长负责查房。

3. 保安

按照每收治200名患者配备1名治安民警（或第三方保安）的标准配备。每两个人为一组，每一班8小时，轮班上岗，主要负责在三区两通道进行守护工作。

4. 保洁

按照每收治100名患者配备4～6名保洁人员的标准进行配备。每4～6个人一组，每一班8小时，轮班上岗。保洁人员轮班进入污染区进行清洁工作。

5. 消防管理人员

每收治 500 名患者配备 1 名消防管理人员。

6. 药师

每收治 100 名患者配备 1 名西药师，每收治 500 名患者配备 1 名中药师。

7. 院感人员

每收治 50 名患者配备 1 名院感人员。

8. 消防人员

每收治 500 名患者配备 1 名消防管理人员。

三、医用防护用品配置

（一）医用防护用品配置品类和质量标准

医用防护用品是保证方舱医院运行的重要物资，包括紧急防护服、N95 口罩、一次性护目镜、防护面罩和一次性医用手套等。防护物资可按照每收治 1000 名患者每天约配备 600 套的标准进行配置。如有条件可配备其他防护物资，包括长筒防污染靴、防水防污染鞋套、外科手术服、外科手术帽和一次性医用床单。由于防护用品供应量紧张，防护用品的物资需求由方舱医院每日报送，由应急及物资保障组下设需求与分配组进行统计，后由采购与调拨组每日按需进行调配。部分主要物资的防护标准如下表所示。

医用防护用品配置品类和质量标准

序号	采购物品	质量特性描述
1	N95 口罩	（1）符合医用防护口罩的技术要求 GB 19083—2010，或欧标 FPP2 型（在结核病治疗中心或实验室 TN 检测中的 FPP3 型） （2）空气里流量每分钟 85 L 条件下： ①颗粒过滤效率：PFE（0. 1 μm）>95% ②细菌过滤效率：BFE（0. 3 μm）>95% ③防护 NaCl 气溶胶 0. 075 μm ④不带呼吸阀
2	护目镜	（1）使用无通风孔的护眼塑料，聚碳酸酯或新丙烯镜片 （2）具有弹性系带 （3）视野广，亮度好，防溅功能好，皮带松紧耐磨
3	一次性外科口罩	（1）符合医用外科口罩的技术要求 YY 0469—2011 （2）使用 3 层医用无纺布 （3）带鼻夹和 4 条绑线/扎带，或弹性耳挂 （4）空气里流量每分钟 85 L 条件下，细菌过滤效率 >95%；符合 EN 14683：2005/Ⅱ或更严格标准

续表

序号	采购物品	质量特性描述
4	一次性医疗防护服/连帽工作服	（1）符合医用防护服 GB 19082—2009 的技术要求 （2）具有加固的接缝和袖口，并在脚踝、面部和腰部有弹性设计 （3）拉链翻盖，带有黏性封口，并有手指封扣设计 （4）符合 EN14126、EN 13982-1（类型 5）、EN 13034（类型 6）、EN 14605（类型 4）等标准，符合欧洲法规或美国同等法规对Ⅲ类个人防护设备的要求，或更严格标准
5	一次性医用乳胶手套	（1）符合医用乳胶手套 GB 10213—2006 的技术要求 （2）100% 腈质材料，不掉粉屑 （3）手腕处不低于 280 mm，直边或卷边设计 （4）或符合 EN 455-1/2/3/4：2001/2007/2009 规范的进口产品

（二）医用防护用品配置数量

医用防护用品配置数量标准（数量以 1000 张床 10 日需求计）

序号	采购物品	单位	数量
1	一次性连体防护服（带鞋套）	套	5000
2	护目镜（可消毒后重复使用）	个	5000
3	一次性使用医用帽	顶	5000
4	医用口罩	个	20 000
5	面屏	个	2000
6	全面型防护面屏	个	200
7	医用防护口罩（N95）	个	5000
8	防渗隔离衣（一次性使用手术衣）	套	5000
9	一次性乳胶手套（检查手套）	双	15 000
10	无菌手套（外科手套）	双	5000
11	一次性长筒鞋套	双	5000

（三）医用防护用品配置测算依据

方舱内医用防护用品的配置数量主要取决于隔离病区患者数量，集中收治场所医护人员数量，卫生专业技术人员数量，以及保安、保洁、配送等辅助工作人员数量。由于需求量巨大，如果每日配送的数量不一定能够满足前期设定的数量标准，需要根据实际情况酌情延长口罩等的使用时间，减缓消耗速率。货源充足时按照标准配置，缺口较大时应灵活机动处理。

此外，由于医用防护物品使用的不确定性，数量计算均为大致估计，无严格计算公式。以每个病区配备的医护人员所需防护服为例。每个病区（50 张床位）相应配备 5 名医生。每天 5 名医生中的 3 名进入舱内污染区进行日常医疗服务，一人一次 8 小时，三班倒，共需 9 套防护服。另外 2 名医生中 1 人充当责任医生，负责每天早晨进舱查房，额外需要 1 套，医生共计 10 套。同时，每个病区配备 20 名护士，每班次进舱 3 人，一天 24 小时 6 班人员交替，共计 18 套，另外每个病区负责查房的护士长 1 套，护士共计 19 套。相应的保安人员、保洁人员等也按照实际到岗的班次和每班人数进行配备，遵循一班一人一套的原则。

在方舱医院运行时，防护物资的补给规则遵照以上的估计原则，结合各家方舱医院每日上报需求，分配与调配组可安排配送 1 ~3 日量至方舱医院。

四、医用设备配置

参考《新型冠状病毒感染的肺炎诊疗方案（试行第七版）》及《方舱医院院内诊疗方案（第一版）》，由于方舱医院主要收治轻症患者，以提供基本治疗、防止疾病扩散同时兼顾方舱患者其他疾病处理为目的，另外为了防止轻症患者突然出现危急情况，在设备配置时应配齐急救设备和抢救药品。应急及物资保障组拟定方舱医院医用设备目录如下表。

医用设备配置

序号	采购物品	单位	数量
1	心电监护仪	台	10
2	除颤仪	台	2
3	呼吸机（无创）	台	5
4	床旁超声	台	1
5	数字十二通道心电图机	台	2
6	电动吸痰器	台	10
7	床单位消毒机	台	5
8	血气分析仪（手持式）	台	1
9	五分类血液分析仪	台	1
10	微量泵	个	6
11	输液架	台	100
12	发药车	台	8
13	护理车	台	8
14	输液车	台	8
15	配药车	台	8
16	病历文书车	台	8
17	抢救车	台	8

续表

序号	采购物品	单位	数量
18	简易呼吸器	件	10
19	0. 5 mmpb 铅帽	件	1
20	铅围脖	件	1
21	铅围裙	件	1
22	铅衣服	件	1
23	不锈钢体温表盘	个	10
24	氧气瓶配套加压阀湿化器	台	50
25	指夹式血氧仪	台	50
26	心肺复苏按压板	块	10
27	转运平车	台	10
28	电子喉镜	个	1
29	B 超机	台	1
30	移动 CT	台	1
31	移动 DR	台	1
32	全自动干式生化仪	台	1
33	标本转运盒	个	10

五、医用耗材配置

参考《新型冠状病毒感染的肺炎诊疗方案（试行第七版）》及《方舱医院院内诊疗方案（第一版）》，结合普通医院病房的耗材消耗，拟定方舱医院医用耗材目录，配置数量以1000张床10日需求计算。在方舱医院运行时，医用耗材的补给应结合各家方舱医院每日上报需求，采购与调拨组可安排配送1～3日量至方舱医院。方舱医院医用耗材配置如下表。

医用耗材配置（数量以1000张床10日需求计）

序号	采购物品	单位	数量
1	氧气瓶	个	50
2	紫外线柜式消毒机	台	10
3	空气消毒机	台	140
4	快速手消毒液	瓶	1000
5	消毒液检测试纸	包	10
6	75% 医用酒精 500 mL	瓶	400
7	0. 5% 碘伏（50 mL）	瓶	100
8	含氯泡腾片	瓶	100
9	一次性使用氧气面罩	个	300
10	普通输液器（自动止液功能）	个	500
11	一次性使用止血带	个	500
12	棉签	袋	500
13	一次性治疗巾（非无菌）	个	100
14	无菌治疗巾	片	100
15	无菌纱布	袋	1000

续表

序号	采购物品	单位	数量
16	无菌棉球	包	1000
17	1 mL 注射器	个	100
18	2 mL 注射器	个	100
19	5 mL 注射器	个	100
20	20 mL 注射器	个	100
21	尿管	根	50
22	导尿包	个	50
23	医用纱布片 7 * 9（不显影）2 片装	包	500
24	医用纱布片 7 * 9（显影）5 片装	包	100
25	一次性使用吸氧管（带湿化液）100 mL	套	50
26	一次性使用吸氧管（带湿化液）250 mL	套	50
27	血糖试纸（12 台血糖仪，同品牌）	片	500
28	心电监护电极	个	200
29	胃管	个	50
30	输液敷贴	个	500
31	一次性使用真空采血管	支	500
32	一次性使用静脉采血针	支	300
33	水银体温计	支	300
34	听诊器	副	100
35	氧气枕	只	50

续表

序号	采购物品	单位	数量
36	移动输液架	个	100
37	治疗盘	个	50
38	弯盘	个	50
39	浸泡桶（消毒弯盘、治疗盘）	个	50
40	轮椅	台	10
41	体重秤	台	5
42	红外体温枪	台	20
43	身高体重秤	台	5
44	电子血压计	台	30
45	观片灯（单联）	台	3
46	观片灯（双联）	台	3

六、医用药品配置

参考《新型冠状病毒感染的肺炎诊疗方案（试行第七版）》《方舱医院院内诊疗方案（第一版）》的推荐目录，以及《国家基本药物目录》和《国家处方集》等，拟定方舱医院配置药品目录，配置数量以 1000 张床 10 日需求计算。针对实际的药品分配情况，应在尽量满足方舱医院的日常药品需求的前提下，结合统筹医院的药品存量，由采购与调拨组组织统筹医院药品采购和分配人员负责对方舱医院进行采购和补给。并保障新冠肺炎轻症患者药物治疗和其他慢性病治疗的用药需求，以及突发危急重症患者的抢救用药保障。方舱医院医用药品配置如下表。

医用药品配置（数量以1000张床10日需求计）

序号	名称	规格	单位	数量
1	盐酸莫西沙星片	0.4 g*3片	盒	1000
2	阿莫西林克拉维酸钾片	0.487 g	盒	200
3	注射用头孢曲松钠	1.0 g	支	200
4	呋塞米注射液（速尿针）	20 mg*2 mL	支	50
5	硫酸阿托品注射液	1 mL：0.5 mg	支	50
6	尼可刹米注射液（可拉明）	375 mg：1.5 mL	支	50
7	去乙酰毛花苷注射液（西地兰）	0.4 mg：2 mL	支	50
8	硝苯地平控释片	30 mg*7片	盒	250
9	硝酸甘油片	0.5 mg*100片	瓶	50
10	盐酸甲氧明注射液	1 mL：10 mg	支	50
11	重酒石酸去甲肾上腺素注射液	2 mg：1 mL	支	50

续表

序号	名称	规格	单位	数量
12	重酒石酸间羟胺注射液（阿拉明）	1 mL：10 mg	支	50
13	盐酸胺碘酮片	0. 2 g	盒	50
14	盐酸多巴胺注射液	2 mL：20 mg	支	50
15	盐酸肾上腺素注射液	1 mg：1 mL	支	50
16	盐酸异丙肾上腺素注射液	1 mg：2 mL	支	50
17	盐酸甲氧氯普胺注射液（胃复安针）	1 mL：10 mg	支	50
18	复方氨基酸注射液(18AA-V)	250 mL：8. 06 g 总氨基酸 + 12. 5 g 木糖醇	支	100
19	板蓝根颗粒	10 g * 15 袋	包	1000
20	地塞米松磷酸钠注射液	1 mL：5 mg	支	100
21	氯化钾缓释片	0. 5 g * 24 粒	盒	150
22	蒙脱石散	3 g * 12 袋	盒	150
23	双氯芬酸钠栓	50 mg * 10 枚	盒	100
24	维生素 B_1 片	10 mg * 100 片	瓶	500
25	维生素 B_6 片	10 mg * 100 片	瓶	500

续表

序号	名称	规格	单位	数量
26	维生素 C 片	100 mg * 100 片	盒	150
27	精蛋白生物合成人胰岛素注射液（诺和灵 30R 笔芯）	300 IU : 3 mL	支	50
28	洛匹那韦/利托那韦片	200 mg/50 mg * 120 片	盒	350
29	磷酸氯喹片	0.25 g * 100 片	盒	200
30	利巴韦林注射液	0.1 g	支	100
31	利巴韦林片	100 mg * 25 片	盒	2000
32	磷酸奥司他韦胶囊	75 mg * 10 粒	盒	500
33	氯化钠注射液（0.9% NaCl）	500 mL : 4.5 g	袋	900
34	0.9% 氯化钠注射液 10 mL	10 mL : 90 mg	支	500
35	金叶败毒颗粒		盒	500
36	蓝芩口服液	10 mL * 6 支	盒	500
37	泰诺	20 片	盒	100
38	铝碳酸镁咀嚼片	0.5 g * 20 片	盒	100
39	多潘立酮片	10 mg * 42 片	盒	100
40	芬必得胶囊		盒	100

续表

序号	名称	规格	单位	数量
41	氯雷他定片	10 mg＊6 片	盒	100
42	小檗碱片		盒	100
43	氨氯地平片	5 mg＊7 片	盒	200
44	美托洛尔缓释片	47.5 mg＊7 片	盒	100
45	氯吡格雷片	75 mg＊7 片	盒	100
46	阿司匹林片	0.1 g＊30 片	盒	100
47	阿托伐他汀	20 mg＊7 片	盒	100
48	二甲双胍片	0.5 g＊20 片	盒	200
49	艾司唑仑片	1 mg＊150 片	盒	10
50	吸入用布地奈德混悬液	1 mg：2 mL（5 支/袋）	袋	50
51	吸入用异丙托溴铵溶液	2 mL：250 μg＊10 支	盒	50
52	氨茶碱缓释片	0.1 g＊24 片	盒	10
53	头孢克洛胶囊	0.25 g＊6 粒＊2 板	盒	50
54	藿香正气软胶囊	0.45 g＊24 粒	盒	3000
55	金花清感颗粒	5 g＊6 袋	盒	2000
56	金莲花软胶囊	0.55 g＊24 粒	盒	2000
57	强力枇杷露	250 mL	盒	2000

续表

序号	名称	规格	单位	数量
58	莲花清瘟胶囊	0.35 g * 24 粒	盒	2000
59	注射用奥美拉唑钠	40 mg	支	200
60	异甘草酸镁注射液	10 mL：50 mg	支	100
61	盐酸消旋山莨菪碱注射液	10 mg * 1 mL	支	100
62	盐酸利多卡因注射液	0.1 g * 5 mL	支	250
63	盐酸阿比多尔片	0.1 g * 12 片	盒	1200
64	维生素 C 注射液	2 mL：0.5 g	支	2000
65	乳酸左氧氟沙星氯化钠注射液	0.2 g * 100 mL	支	100
66	葡萄糖注射液(50% 浓糖)	20 mL：10 g	袋	500
67	葡萄糖注射液(5%)	500 mL：25 g	袋	500
68	葡萄糖注射液(5%)	250 mL：12.5 g	袋	500
69	葡萄糖注射液(5%)	100 mL：5 g	袋	500
70	葡萄糖注射液(10%)	250 mL：25 g	袋	500

续表

序号	名称	规格	单位	数量
71	葡萄糖氯化钠注射液	500 mL：25 g(5%)	袋	500
72	氯化钠注射液(0.9% NaCl)	250 mL：2.25 g	袋	500
73	氯化钠注射液(0.9% NaCl)	100 mL：0.9 g	袋	500
74	洛贝林	3 mg：1 mL	支	250
75	复方甘草酸苷注射液	20 mL	支	100
76	呋塞米片（速尿片）	20 mg * 100 片	瓶	100
77	唑吡坦片	10 mg * 20 片	盒	50
78	吲哚美辛栓	0.1 g * 10 粒	盒	100
79	银黄含片	0.65 g * 24 片	盒	200
80	盐酸西替利嗪片	10 mg * 6 片	盒	50
81	头孢地尼	0.1 g * 10 片	盒	150
82	舒利迭	250 μg：50 μg * 60 吸（泡）	盒	100
83	口服补液盐Ⅲ	5.125 g * 6 包	盒	300
84	小檗碱	0.1 g * 100 片	瓶	50
85	甘草合剂	100 mL	瓶	500
86	复方甲氧那明胶囊	60 粒	瓶	100
87	对乙酰氨基酚	0.3 g * 24 片	盒	100

续表

序号	名称	规格	单位	数量
88	地西泮注射液	2 mL : 10 mg * 10 支	盒	150
89	丙戊酸钠	0.2 g * 100 片	瓶	100
90	美托洛尔	25 mg * 20 片	盒	50
91	阿卡波糖片（拜唐苹）	50 mg * 30 片	盒	100
92	整肠生		盒	100
93	开塞露		支	300
94	疏风解毒胶囊/颗粒	粒	盒	2000
95	清肺排毒汤		袋	按实际配送
96	1 号汤剂		袋	按实际配送
97	2 号汤剂		袋	按实际配送
98	3 号汤剂		袋	按实际配送

九州通医药集团股份有限公司物流公司入驻武汉国际博览中心红十字会防控物资仓库，负责物资的物流管理环节

第三章　方舱医院人员与医用物资管理

一、人员管理

（一）方舱医院患者管理

方舱医院作为应对严重疫情而迅速建立的大规模临时医疗场所，能够及时提供大容量、低成本的简易医疗床位，为大量确诊轻症患者，提供及时有效的治疗和隔离。本小节主要介绍方舱医院患者管理中 3 项重要的标准：收治、转诊和出舱标准。

1. 收治标准

参考武汉市新型冠状病毒肺炎防控指挥部医疗救治组制定的《方舱医院管理规则》，方舱医院原则上收治已确诊并且未在定点医院隔离治疗的轻症患者，具体收治的患者应满足如下条件：

①新型冠状病毒肺炎的确诊病例。

②对应国家卫生健康委发布的《新型冠状病毒肺炎

诊疗方案（试行第七版）》中临床分型为轻型或普通型。

③有生活自理能力，且年龄在 18～65 岁。

④无呼吸系统、心脑血管系统相关的基础性疾病，且无精神疾病。

⑤流感病毒核酸检测为阴性。

2. 转诊标准

如患者出现《新型冠状病毒肺炎诊疗方案（试行第七版）》中临床分型为重型或危重型患者，应及时转运定点医院进行治疗，具体标准摘录如下。

（1）临床分型为重型患者

成人符合下列任何一条：

①出现气促，RR≥30 次/分。

②静息状态下，指氧饱和度≤93%。

③动脉血氧分压（PaO_2）/吸氧浓度（FiO_2）≥300 mmHg（1 mmHg＝0.133 kPa）。高海拔（海拔超过 1000 米）地区应根据以下公式对 PaO_2/FiO_2 进行校正：PaO_2/FiO_2×[大气压(mmHg)/760]。

④肺部影像学显示 24～48 小时内病灶明显进展＞50% 者。

儿童符合下列任何一条：

①出现气促（＜2 月龄，RR≥60 次/分；2～12 月龄，RR≥50 次/分；1～5 岁，RR≥40 次/分；＞5 岁，RR≥30 次/分），除外发热和哭闹的影响。

②静息状态下，指氧饱和度≤92%。

③辅助呼吸（呻吟、鼻翼扇动、三凹征），发绀，间歇性呼吸暂停。

④出现嗜睡、惊厥。

⑤拒食或喂养困难，有脱水征。

（2）临床分型为危重型患者

符合以下情况之一者：

①出现呼吸衰竭，且需要机械通气。

②出现休克。

③合并其他器官功能衰竭需 ICU 监护治疗。

另外，如患者出现并发症或伴随疾病达到需要转综合医院治疗的情况，应转运定点综合医院进行治疗。

3. 出舱标准

按照《新型冠状病毒肺炎诊疗方案（试行第七版）》中规定，患者出舱应该满足如下条件：

①体温恢复正常 3 天以上。

②呼吸道症状明显好转。

③肺部影像学显示急性渗出性病变明显改善。

④连续两次痰、鼻咽拭子等呼吸道标本核酸检测阴性（采样时间至少间隔 24 小时）。

患者出舱后应遵照《新型冠状病毒肺炎诊疗方案（试行第七版）》中出院后相关注意事项，完成隔离管理和有效随访、复诊。

（二）方舱医院医护与辅助工作人员管理

1. 人员培训

（1）对在方舱医院的全体工作人员进行统一的基础培训

内容包括该传染病相关院感防控知识的培训和对方舱医院整体布局的培训。可采用现场培训和网络授课考核相结合的多样化形式，督促全体工作人员及时完成学习培训，确保熟练掌握方舱医院各岗位具体的职责和工作流程，熟悉方舱医院的平面布局及“三区两通道”的划分，重点了解该目标传染病感染的防控知识、方法与技能，在高效完成方舱医院工作任务的同时做到有效防控，降低职业暴露风险，切实保护自身和方舱中其他医务人员的安全。

（2）依据不同岗位人员制订不同的进阶培训内容

对在方舱工作的医护人员的进阶培训内容包括：前期，熟悉方舱医院所使用医疗设备的操作手册、使用规范和消毒流程；在与病患的接触过程中，如何保证在规范治疗和用药的前提下，对患者的心理和不良情绪进行安抚和疏导；在防护方面，对可能参加高风险操作（如可能与确诊或疑似患者的接触等）或可能进入半污染区和污染区的医护人员要重点培训和考核。

中医药在新冠肺炎疫情防控和救治中发挥了
重要作用（国药控股湖北有限公司供稿）

2. 关注方舱医护人员健康

面对巨大的工作量和院感防护的双重压力，应合理调配人力资源和科学排班以避免医护人员过度劳累，同时也要注意日常合理饮食以保证正常的免疫力。在每日工作前和结束工作后，开展主动健康监测，包括体温和呼吸系统症状等。以下情况应格外注意，做好防治工作：①对于一线工作人员，在穿上防护用品后会出现不透气和大量出汗的情况，须防止因低钾、低血糖引起的心脏骤停等疾病，可多饮用含钾饮料，如果汁；②由于过度繁忙和高度紧张，医护人员可能会出现血压骤升等情况，如医护人员患有脑血管疾病，可能造成脑血管破裂等严重现象，应主动监测血压，并常备钙抗剂和β受体拮抗剂等药物；③对于伴有睡眠呼吸障碍性疾病的医护人员，即伴有呼吸暂停现象的打鼾者，应尽量在睡眠中佩戴CPAP呼吸机，防止因过度疲劳造成夜间严重缺氧而诱发的危急并发症。

由于方舱医护人员工作场所及岗位的特殊性，还应关注其心理健康及情绪管理。轮流值班时，需保证适宜的时间间隔，保证合适的休息和放松，保证充足的睡眠，建议保证正常的休闲娱乐节奏，尽量保持以往的作息时间。必要时评估医护人员的心理状况，根据医护人员心理防护指南及原则对有需要的医护人员进行心理疏导。在有条件的情况下配备心理治疗师，或让有心理治疗资质人员为医护和辅助人员进行心理辅导。医护人员是方舱医院抗疫前线重要的一员，应采取多种措施，保障其身心健康，才能为患者提供优质的医疗服务。

二、医用物资管理

（一）普通医用物资的管理

新冠肺炎疫情防治期间，为切实做好方舱医院医用物资管理工作，确保设备、药品和耗材的质量安全，根据国家卫生健康委《新型冠状病毒肺炎诊疗方案（试行第七版）》《医疗器械监督管理条例》《国家基本药物目录（2018 版）》等相关法律法规规定，结合方舱医院实际情况，疫情期间方舱医院普通医用物资管理工作参考以下制度。

1. 方舱医院配置标准的制定

①方舱医院人员和物资配置标准由应急及物资保障组领导小组牵头制定，配置标准综合考虑普通传染病房和重症监护病房的人员配置与常用设备配置标准。

②方舱医院配置标准的确定。方舱医院应急及物资保障组领导小组按照国家卫生健康委发布的《新型冠状病毒肺炎诊疗方案（试行第七版）》《方舱医院院内诊疗方案（第一版）》等相关规定，会同方舱医院医务管理部门及临床专家意见，以 1000 张床位为基准确定方舱内服务人员的数量，并依据以上人员配置拟定防护用品配置数量。如有方舱医院有特殊情况需要调整配置标准的，应由方舱管理人员综合方舱内临床专家（医护人员）和主要职能负责人（院感、治安、保洁）意见，报送应急及物资保障组物资需求与分配组，经专家审核后临时更改该医院配置。

2. 医用物资采购制度

①方舱医院医用物资的采购由应急及物资保障组下

设的需求与分配组负责制订相应目录，制订采购和请领计划，由采购与调拨组统一采购，并根据实际情况和疫情进展及时更新产品采购目录。

②医用物资，特别是防护物资，需要严格按照疫情一线的应用场景进行采购，应将详细的物资产品型号和防护、应用等标准向供应厂商进行公布。

③采购数量原则上应首先满足《关于调整方舱医院医用物资开舱配置标准的通知》中的方舱医院开舱物资标准，在运行时针对防护用品、医用耗材和药品，采购量应尽量满足方舱医院（满床位）10 日的需求。

④对于医用防护物资、医用设备与医用耗材的采购，必须从有资质的正规经营企业购入，应将有关业务关系的经营企业和业务人员的资质备案。

⑤以上物资的采购原则上可不经过招标流程，但相应的订单目录在讨论过程中应有外来临床或医疗机械、耗材专家参与，形成书面文档统一留档记录。

⑥对于药品的采购，应由需求与分配组和药品保障供应相关负责人制订相应药品目录，制订采购和请领计划，由采购与调拨组统一采购。

⑦药品采购计划及品种的确定，应按照国家卫生健康委发布的《新型冠状病毒肺炎诊疗方案（试行第七版）》《方舱医院院内诊疗方案（第一版）》《国家基本药物目录》《国家处方集》等，会同方舱医院医务管理部门及临床专家意见，确定采购药品名单，并根据治疗

进展及时更新药品目录。超出医院处方集的药品，牵头医院应根据《临时采购药品管理规定》，启动临时采购药品程序。

⑧药品必须从有资质的正规药品经营企业进行采购，应将有关业务关系的经营企业和业务人员的资质备案。

3. 医用物资验收制度

①方舱医院医用物资的验收由应急及物资保障组下设的管理审核组负责，对采购产品进行质量验收，合格后方可投入使用。

②验收时，应对医用物资，特别是防护物资的具体型号进行核对，对于不符合要求的产品绝不验收和入库，并不予准许从厂商发货到方舱医院。

③验收时必须有发票或随货同行单，对于防护用品、医疗设备、医用耗材和药品须有医疗器械注册证或检验报告书等合格认证，进口药品须有《进口药品注册证》及《进口药品检验报告书》，相关的认证文件应加盖供货单位公章。

④应对购入物资的数量、包装、标签说明书、注册商标、批准文号、批号效期、检验报告等项目进行验收，对数量或质量有疑问的拒绝接收，应与供货单位联系，做相应处理。

⑤应建立完整的物资购进验收记录，注明产品名

称、供货单位、规格、单位、数量、生产厂家、批号、效期、包装、批准文号，并对购货数量、价格、入库日期、验收结果等项目进行记录，由验收人签名。

⑥针对有效期进行合理管控，有效期为 1 年的产品，效期剩余 3 个月不得验收，有效期为 1 年以上的产品，效期剩余 6 个月不得验收。验收人员验货时发现近效期的产品应立即通知供货单位退货，如遇临床必需，采购较困难的产品，可酌情放宽效期范围，并向方舱医院应急及物资保障组领导小组进行书面汇报后再确定是否验收。

4. 医用物资的分配与调拨制度

①方舱医院医用物资的验收由应急及物资保障组下设的分配与调拨组负责，对采购物资进行分配和调拨。

②在方舱医院前期建设阶段，应根据方舱医院的拟开设床位数，依据《关于调整方舱医院医用物资开舱配置标准的通知》中的配置标准，调集相匹配的医用物资以满足医院开舱条件。

③在方舱医院运行阶段，应由需求与分配组对接方舱医院每日上报的物资需求量，统计物资缺口，并由采购与调拨组进行调配。

④应急及物资保障组下属各组应配合保证存量，当需要对方舱医院进行物资补充，可从特定仓库中对医院进行物资运送，如遇时间紧迫需要供应厂商直接发货的

情况，管理审核组应派遣专人到指定方舱医院进行验收与监督。

南京医药湖北有限公司在行动

（二）捐赠医用物资的管理

新冠肺炎疫情防治期间，为切实做好疫情相关的捐赠医用物资监管工作，确保捐赠医用物资符合标准，且质量安全，根据原卫生部《医疗卫生机构接受社会捐赠资助管理暂行办法》等相关法律法规，参考《关于加强捐赠救灾药品和医疗器械监管工作的通知》，结合方舱医院工作实际，现就疫情期间方舱医院捐赠医用物资管理工作制定如下制度。

1. 捐赠物资管理组织

捐赠物资相关管理事宜由方舱医院应急及物资保障组领导小组统筹规划。需求与分配组主要负责抗疫捐赠物资管理工作，建立相关台账；管理审核组负责对捐赠物资的质量进行检验核查；采购与调拨组负责捐赠物资的调剂、发放工作；并由专人向临床专家或药事相关负责人核证物资的临床论证与评估。各小组应做好相关记录，方便后期查询及配合审计、监督工作，并定期向应急及物资保障组领导小组汇报工作动态。

2. 接受捐赠物资的原则

①接受社会捐赠资助必须遵守国家法律法规，捐赠资助要坚持自愿无偿原则，符合公益目的。

②遵循尊重捐赠者意愿与符合医院利益相统一，并符合新型冠状病毒肺炎疫情防治需求的原则。

③不得损害公共利益和公民的合法权益，不得接受附有影响公平竞争条件的捐赠资助，并根据新型冠状病毒肺炎疫情需要，接受符合国家相关质量标准、环保等要求的合格物资（能提供合格的产品证明）。

3. 接受捐赠物资的标准

①境内生产的物资，必须是由国家相关机构批准生产、获得批准文号且符合质量标准的品种，有效期距失

效日期必须在6个月以上。

②境外生产的物资，应是在注册国（地区）合法生产并上市且符合质量标准的品种，符合防护、医用标准的物资，且有效期距失效日期必须在6个月以上。

4. 接受捐赠物资的流程管理

①制订方舱医院捐赠物资目录

根据国家卫生健康委和国家中医药管理局发布的新型冠状病毒肺炎诊疗方案中的推荐用药，以及针对新型冠状病毒肺炎的院感和防护标准，制订接受捐赠物资目录，并定期更新，目录内的物资可直接接受捐赠。

②非目录内捐赠物资处理流程

需求与分配组接到捐赠方捐赠意愿后，应明确捐赠方提供的产品资料、说明书等材料，如是非目录内捐赠物资，应征询临床医疗团队和临床专家意见，确认产品进行临床论证及评估，确定是否接受捐赠。

③接受捐赠物资的数量

依据物资用法用量、疫情时间、疫情发展阶段、采购物资是否满足方舱医院需求等方面，由领导小组酌情确定接受捐赠物资的数量。

5. 捐赠物资的验收

管理审核组验收时实行双人验收制度。验收时需要

上药科园信海医药荆州有限公司捐赠药品

对产品外包装、名称、规格、数量、生产批号、有效期、生产厂家、批准文号（或进口注册证）、质检报告逐一核对。

6. 捐赠物资的储存保管

专库存放、严格按照物资存放要求分类存放，专人管理。做好库房温湿度管理工作，确保耗材、药品等储存安全。做好耗材、药品养护记录及效期管理。

7. 捐赠物资的账物管理

建立捐赠物资专用入库账目和领用登记账目。定期做好捐赠物资盘点，保证账物相符。

8. 捐赠物资的监督

主动接受监督，确保捐赠工作规范有序。物资库房定期对所接收的捐赠物资入库明细和捐赠物资领用登记在相应平台公示。

三、应急管理

（一）信息系统故障管理

信息系统是管理方舱医院的必要手段，关系到舱内患者的诊断、治疗、核查、药品和耗材的统计与管理等诸项重要事项，一旦方舱医院发生信息系统故障，应及时处理，并遵循以下原则：

①立即通知后勤信息系统工程师，寻找原因并及时检修。

②及时上报方舱医院应急及物资保障组，各项工作统一听从安排，以防止发生差错事故。

③系统故障能快速排查后解决的，系统正常后继续工作。

④长时间不能解决的，信息记录流程转入手工工作

模式，期间的处方、耗材分配等工作应及时记录，任何单据应完整保留照片，以备网络正常后的数据核查和补录。

（二）消防应急管理

方舱内人员应提前熟悉方舱医院内消防通道所在位置及消防设施所在位置和使用方法，并进行消防培训。

如遇火情，应该迅速关闭工作场所设备电源开关，并立即向方舱管理人员汇报。如火情无法控制，在保证人身安全的前提下，迅速疏散患者，如有条件应按贵重设备、药品先转运的原则搬离设备及药品，有组织、有秩序地迅速撤离火灾现场。

如遇水淹，应迅速联系医院后勤负责人或相关人员，组织抢修，关闭泛水管道阀门，阻断水源蔓延。尽可能及时地清扫污水，在保证人身安全的前提下，对信息系统硬件及时断电，保护信息资料，避免资料丢失，并尽快将贵重、特殊药品移到高位，避免被水淹没，如有条件可抢运各类物资设备，把损失减至最低限度。

（三）停电应急管理

由于方舱医院中储存有患者日常所需的医用耗材和药品，部分该类物资需要冰箱冷藏，如遇突发停电情况，应当在做好舱内患者安抚工作的同时，立即关闭各重要仪器设备的开关，上报方舱医院后勤部门，通知电工维修。在医护人员和药事人员的指导下尽量保证医用耗材和药品安全。如有条件，应记录冰箱温度，合理转移要求低温储存的药品，减少损失。在停电期间，安保人员应在做好个人疾病防护的同时加强巡视，做好防火防盗工作。

附　　录

一、相关政策法规

《湖北省方舱医院救治工作管理规范》

《新型冠状病毒感染的肺炎疫情紧急心理危机干预指导原则》

《新型冠状病毒感染的肺炎防控中常见医用防护用品使用范围指引（试行）》

《新型冠状病毒肺炎诊疗方案（试行第七版）》

《中华人民共和国药品管理法》

《中华人民共和国药品管理法实施条例》

《医疗卫生机构接受社会捐赠资助管理暂行办法》

《关于加强捐赠救灾药品和医疗器械监管工作的通知》

《药物临床试验质量管理规范》

《医疗机构药事管理规定》

《方舱医院院内诊疗方案（第一版）》

《临时采购药品管理规定》

《处方管理办法》

《冠状病毒 SARS-CoV-2 感染：医院药学工作指导与防控策略专家共识（修订版）》

二、《临时改建方舱医院消防安全要求（试行）》

（一）用于临时改建为集中收治新型冠状病毒感染的肺炎患者方舱医院（以下简称“方舱医院”）的体育馆、展览馆、校舍建筑，耐火等级不应低于二级，方舱医院宜设置在建筑首层及二层，不得设置在建筑的地下或半地下。

（二）方舱医院每层平面宜划分为病区，每个病区设置床位不宜超过 100 个，病区之间宜设置实体墙或具备同等耐火等级的轻质防火隔墙进行防火分隔，墙体高度应至楼板底，高大空间设置确有困难的，其墙体高度不宜小于 2 米，隔墙上的门应采用乙级防火门。

（三）方舱医院每个病区床位布置应划分护理单元，每个护理单元床位宜设 30 ~ 40 张，护理单元之间应按井字形布局，护理单元之间的疏散通道宽度不应小于 1.4 米，正对安全出口的疏散通道宽度不应小于 2.0 米。

（四）方舱医院应充分利用建筑既有疏散楼梯、安全出口等疏散设施，每个病区容纳床位数在 500 个及以下的，安全出口数量不宜少于 2 个，每个病区容纳床位数在 500 ~ 1000 个（含 1000）的，安全出口数量不宜少于 4 个。

（五）方舱医院使用的酒精、强氧化剂等易燃易爆物品应单独存放，存放场所与周围区域宜采用实体防火隔墙或相应耐火等级的轻质墙体、乙级以上防火门窗进行严密分隔。

（六）储存医用氧气的供氧设备应远离热源、火源和易燃易爆源。采用液氧供氧方式时，大于500升的液氧罐应设置在室外，室外液氧罐与病区的距离应大于7.5米；医用氧气管道不应与电线电缆、可燃气体管道敷设在同一管道井内或支架上，且应设置静电接地装置。敷设有医用氧气管道的管道井宜有良好通风。

（七）医疗、生活用电气线路宜分类穿管敷设；具备条件的，宜在新增设供电线路的末端配电箱内增设具有独立报警功能的电气火灾监控装置。

（八）方舱医院病区应使用符合国家标准的合格电器产品，电热毯、插座必须采用通过国家强制性认证（3C）的产品。

（九）用于改建为方舱医院的建筑应确保消防水源充足、消防设施供电安全可靠，室内消火栓系统、自动喷水灭火系统、防排烟系统及火灾自动报警系统应保持完好有效。方舱医院应按严重危险级场所配置灭火器，宜选用水基型灭火器。

（十）方舱医院应强化应急照明和灯光疏散指示标志设置：安全出口、楼梯间和主要疏散通道部位的应急照明设施连续供电时间宜大于1小时，地面最低水平照度不宜低于10 Lx。

（十一）每个护理单元内宜为医护人员配备1～2具过滤式消防自救呼吸器，自救呼吸器应放置在护理区醒目且便于取用的位置。

（十二）方舱医院投入使用的同时，应同步建立“医院—病区—护理单元”三级消防安全管理机制，细化明确各级消防安全管理和应急处置职责。

（十三）方舱医院应严格落实消防控制室24小时双人不间断值班制度；严格落实每2小时病区重点部位防火巡查制度，将安全出口、疏散通道、供氧设施、酒精等易燃医疗物品存放点、病区电器产品、线路桥架管井、变配电室和配电间、消防泵房纳入巡查范围，及时纠正消防安全违法行为；严格落实建筑消防设施检查制度，宜每周进行一次各类消防设施单项功能测试，明确专人承担消防设施维护保养职责，及时做好故障排查、消除工作。

（十四）方舱医院应加强用火用电安全管理，病区内严禁吸烟和使用明火，严禁使用与医疗用途无关的大功率电器。防火巡查时应核查建筑整体用电负荷情况，避免超负荷用电。

（十五）方舱医院应根据所在建筑特点和消防安全重点部位，制定有针对性的灭火和应急预案。医护人员、管理人员、保安人员应接受消防安全培训，掌握基本的防灭火知识及灭火和应急疏散预案的内容、操作程序，具备扑救初起火灾和引导协助患者疏散的能力。

Handbook for the Construction of Makeshift Hospitals

Introduction

Recently, confirmed cases of the 2019 novel coronavirus (2019-nCoV) pneumonia have appeared in a long list of countries and regions with a high growth rate. Since the first report of 4 cases in Wuhan, Hubei Province, China, on December 29, 2019, within a month, 7,736 cases had been confirmed in China, with 12,167 suspected, 1,370 severe, and 170 deaths (World Health Organization report, January 30, 2020)[①].

In the epicenter of the outbreak, Wuhan, the severe shortage of medical resources had resulted in a large number of undiagnosed and untreated patients. Failed to be timely isolated by designated hospitals, these patients were forced to become the primary source of infection. Both prevention and treatment work went extremely difficult.

In response, starting on February 3, 2020, Wuhan began to build a series of makeshift hospitals to treat patients

---

① https: //www. who. int/docs/default-source/coronaviruse/situation-reports/20200130-sitrep-10-ncov. pdf? sfvrsn = d0b2e480_2.

with mild coronavirus infections. The makeshift hospitals originated from self-contained transportable field hospitals brought by the military to the battlefield. In Wuhan, selected sports centers and exhibition centers were reconstructed to form makeshift hospitals, because they had ample space and were flexible enough to adapt to different emergency requirements. The activation of makeshift hospitals had effectively enhanced the city's ability to admit patients and offer medical treatment. As of March 9, a total of 16 makeshift hospitals had been put into use, which had admitted 13,640 patients, of which approximately 12,600 had been discharged.

In this book, we aim to share our experience in managing makeshift hospitals with the countries and regions that are currently suffering from the epidemic. The remainder of this book proceeds as follows. Chapter 1 describes the preparation work for the opening of makeshift hospitals. Chapter 2 elaborates on the allocation of personnel and medical supplies in the makeshift hospital. Chapter 3 summarizes the management of the personnel and medical supplies in the makeshift hospitals.

1 Preparations

1.1 Organizational Structure

1.1.1 Leadership

The leading group—consisting of officers from the National Health Commission of China, State-owned Assets Supervision and Administration Commission of the State Council, and Public Security Bureau of China—was in charge of the management of personnel and medical supplies in all the makeshift hospitals. The group's responsibility included calculating the demand for medical protective equipment, medical devices, medications, and consumables, formulating procurement plans accordingly, and completing medical supplies rationing in the makeshift hospitals. In addition, each makeshift hospital was assigned an acting president who was in charge of the daily management of the hospital. The presidents were healthcare experts who had rich experience in hospital management.

1.1.2 Operating Components

To fulfill its responsibility, the leading group had established four operating components: the demand calculation team, the procurement team, the finance team, and the review team. All teams worked 24/7 to guarantee a timely response to emergencies during the 2019-nCoV outbreak.

(1) The Demand Calculation Team

This team was responsible for estimating demand for medical supplies, including protective supplies, medications, medical equipment, disinfection and sterilization supplies, etc. Based on the statistical analysis, the team would formulate plans to guide the purchase and allocation of medical supplies for each makeshift hospital before and after its activation.

(2) The Procurement Team

The procurement team was responsible for the purchase, the delivery, and the installation of the medial supplies.

(3) The Finance Team

The finance team was in charge of the payment as well as keeping accounting records of the purchased and donated medical supplies.

(4) The Review Team

The team's primary function was quality control. Its responsibility was to ensure that all the medical supplies met the national regulatory requirements.

Personnel Deployment

1.2 Construction of Makeshift Hospitals

Being able to instantly provide substantial low-cost beds to all diagnosed patients with mild symptoms, makeshift hospitals had the potential to reduce the pressure on designated hospitals, isolate the source of infection, and eventually prevent the spread of the epidemic. Guidelines for building a makeshift hospital were given as follows.

1.2.1 Area

A makeshift hospital had to be placed in a building:

—that was relatively isolated, sufficiently far from surrounding residential areas to avoid the pollution of medical waste;

—that was equipped with essential utilities including water and electricity to ensure the necessary living conditions of the patients;

—with wastewater discharge;

—with at least two entrances;

—with the floor space per layer ≥ 10,000 square meters.

10 sport centers, 4 exhibition halls, 2 colleges and universities, and 1 industrial park were selected for transformation. Sports centers were the first choice, not only

because the reconstruction was convenient and cost-effective, but also because they were spacious and contained a large number of entrances and exits that met the requirements of hospital infection control.

1.2.2 Layout

A typical makeshift hospital consisted of an internal area and an external area. Figure 1 illustrated the Wuhan International Conference and Exhibition Centre Makeshift Hospital. The internal area was a ward unit that accommodated patients. The external area comprised a mobile technical support unit and a mobile medical function unit that contained essential diagnostic and treatment departments such as emergency, imaging, surgical room, etc.

A makeshift hospital should meet the standard health and safety requirements. More importantly, the ward unit should consist of three medical sectors—contaminated, semi-contaminated, and clean—and at least two separate passageways—contaminated and clean.

Three sectors:

—Patients in the contaminated section, with 50 beds in one ward;

—The semi-contaminated sector was a transition area

for the staff working in the contaminated sector to perform sterilization before entering the clean sector;

—A separation between different sectors was required to reduce virus transmission;

—Sterilization services were provided regularly in all sectors;

—Recreation areas.

Two passageways:

—Contaminated passageway for patients and clean passageway for staff;

—Securities guarded at the entrance.

1. 2. 3 Number of Beds

The number of beds varied from 400 to 1,600, depended on the size of the hospital:

—Suggested area/volume for one bed was 6 square meters (including passages);

—10,000 square meters per 1,000 patients, 60% of which was for patients, and the rest was for staff and storage.

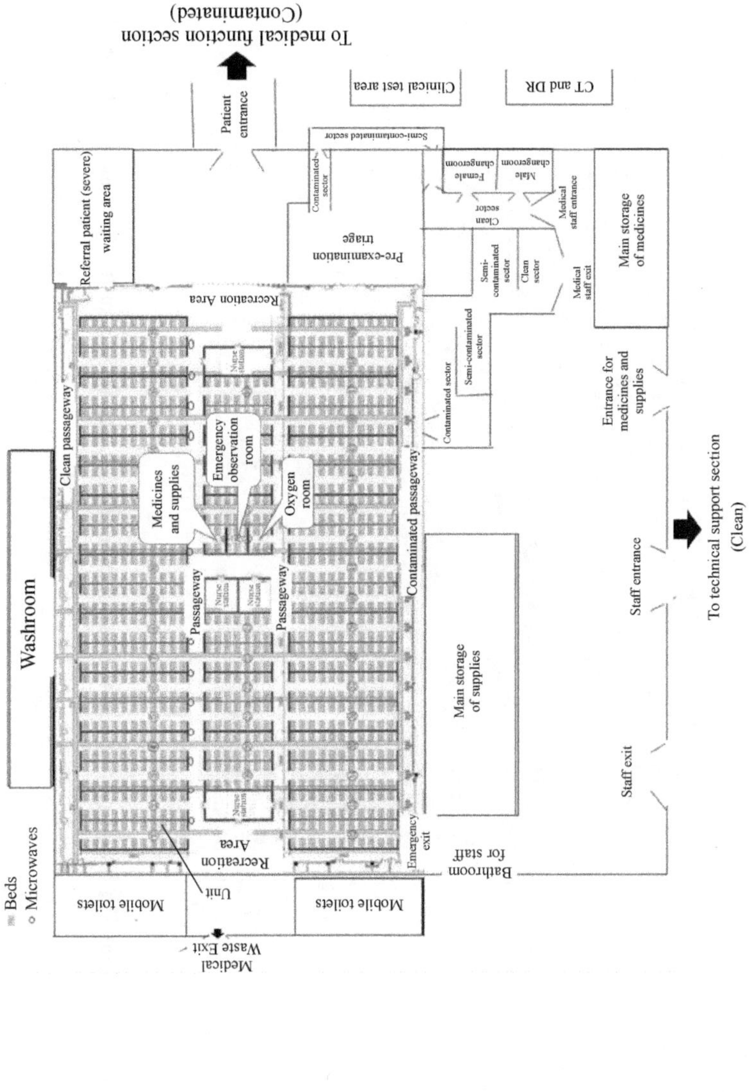

(a) The layout of the internal area

(b) A snapshot of the external area

The Wuhan International Conference and Exhibition Centre Makeshift Hospital

2 Staffing and Medical Supplies Allocation

The Center for Epidemic Prevention and Control in Wuhan issued the *Notice on Adjusting the Standards of Medical Supplies* for the opening of makeshift hospitals on February 20, 2020. The notice determined the standards of personnel and medical supplies in the makeshift hospitals based on 1,000 beds. Hospitals with 500 - 1,000 beds should also follow the rule. The implementation was more flexible if the hospital contained less than 500 beds as long as the hospital could cater to the needs of patients and staff.

A daily supply for the staff was approximately 600 sets for every 1,000 patients admitted. For each hospital, 11 categories of protective supplies, 46 categories of medical consumables, 98 categories of medications (10-day dosage) and 33 types of medical devices were prepared. The allocation of medical supplies followed *The New Coronavirus Infected Pneumonia Diagnosis and Treatment Plan* (*The Seventh Edition, Tentative*), *The Makeshift Hospital In-*

Hospital Diagnosis and Treatment Plan (*The First Edition*), *The National Essential Medicines List*, and *The National Formulary of China*.

2.1 Staff

In principle, the required number of staff depended on the number of admitted patients.

Number of medical staff for 100 patients:

—10 doctors;

—40 nurses;

—1 pharmacist;

—1 traditional Chinese medicine pharmacist (1 more for each additional 500 patients);

—2 sterilization staff.

Number of other staff for 100 patients:

—1 security (1 more for each additional 200 patients);

—1 fire staff (1 more for each additional 500 patients);

—4 –6 cleaning staff.

2.2 Calculation Basis and Allocation Rules

2.2.1 Staff Scheduling

There were multiple wards in the makeshift hospitals, each containing 50 beds.

(1) Doctors

Each ward was assigned 5 doctors.

—3 of the five doctors entered the contaminated area for regular medical services every day (8-hour shift * 3);

—1 doctor in charge of the morning ward round and served as the leader of the 5-people medical team;

—1 doctor stood by;

—The five doctors rotated their roles every day (24 hours).

(2) Nurses

Each ward was assigned 20 nurses.

—18 nurses were divided into groups of three and entered the contaminated area to provide regular nursing services every day (4-hour shift per group * 6);

—2 nurses provided logistics support in the clean area;

—1 head nurse in charge of the ward round;

—The nurses rotated their roles every day (24 hours).

(3) Securities

Every 200 patients were assigned 1 security. In groups of two, the securities took turns to guard the sectors and entrances 24 hours (8-hour shift * 3).

(4) Cleaning Staff

Every 100 patients were assigned 4 – 6 cleaning staff. They took turns to complete the cleaning work in the contaminated area (8-hour shift * 3).

2. 2. 2 Medical Protective Equipment

(1) Quality Specifications

Quality specifications of selected medical protective equipment were as follows.

N95 mask:

—Meet the technical requirements of protective masks for medical use GB 19083—2020;

—Equivalent to FPP2 (FPP3 used in the tuberculosis treatment center, or lab handling tests for TN);

—At an airflow rate of 85 L/min:

Particle filtration efficiency: PPE (0. 1 micron) >95%;

Bacterial filtration efficiency: BPE (0. 3 micron) >95%;

Prevent 0. 075 μm NaCl aerosol;

Without valves.

Goggles:

—Unvented plastic for eye protection;

—Polycarbonate lens;

—Elastic strap;

—Wide field of vision, good brightness, and good anti-splash function.

Disposable surgical mask:

—Meet the technical requirements of protective masks for medical use YY 0469—2011;

—Equivalent to EN 14683: 2005 or more stringent standards;

—3 non-woven fabric layers;

—With nose clip and 4 strings/ties (elastic earloop);

—At an airflow rate of 85 L/min:

Bacterial filtration efficiency: BPE (0.3 micron) >95%.

Disposable medical protective clothing/Coveralls with hoods:

—Meet the technical requirements of protective masks for medical use GB 19082—2009;

—Equivalent to EN14126, EN13982-1 (type 5), EN13034 (type 6), EN14605 (type 4);

—With over-taped seams and cuffs;

—Elastic design for ankle, facial and waist;

—Zip flap with adhesive closure and finger loop.

Disposable medical latex gloves:

—Meet the technical requirements of protective masks for medical use GB 10213—2006;

—Equivalent to EN455-1/2/3/4: 2001/2007/2009; classification according to Directives 93/EC class I;

—100% nitrile, non-powdered materials, not sterile;

—Cuff size≥280 mm, with a straight or rolled edge.

(2) Quantities

Quantity Standard of Medical Protective Equipment

(Based on 10-day usage of 1000 patients)

No.	Items	Quantities
1	Disposable medical protective clothing (with shoe covers)	5,000
2	Goggle (reuse after sterilization)	5,000
3	Disposable medical cap	5,000
4	Disposable surgical mask	20,000
5	Face shield	2,000
6	Full facepiece reusable respirator	200
7	N95 mask	5,000
8	Disposable isolation gown	5,000
9	Disposable medical latex gloves (pair)	15,000

Continued

No.	Items	Quantities
10	Sterile gloves (pair)	5,000
11	Disposable long-tube shoe covers (pair)	5,000

(3) Calculation Basis

The daily demand of each ward was calculated as follows[①]:

—10 sets for 5 doctors: 3 sets/shift * 3 shifts +1 spare set;

—19 sets for 20 nurses: 6 sets/shift * 3 shifts + 1 spare set;

—For security and cleaning staff, follow the rule of 1 set per person per shift.

2.2.3 Medical Device

Quantity Standard of Medical apparatus and instruments

No.	Items	Quantities
1	ECG monitor	10
2	Defibrillator	2

① Due to the high demand, daily delivery might fail to meet the quantity standard. With a severe shortage, the replacement of medical protective equipment should be more flexible.

Continued

No.	Items	Quantities
3	Noninvasive ventilator	5
4	Bedside ultrasound	1
5	Digital 12-channel electrocardiograph	2
6	Electric sputum aspirator	10
7	Bed sterilizer	5
8	Blood gas analyzer (handheld)	1
9	Five-category blood cell analyzer	1
10	Micropump	6
11	Infusion stand	100
12	Medications distributing cart	8
13	Nursing cart	8
14	Infusion cart	8
15	Medications dispensing cart	8
16	Medical record cart	8
17	Medical rescue cart	8
18	Simple respirator	10
19	0. 5 mmpb lead cap	1
20	Lead collar	1
21	Lead apron	1
22	Lead clothing	1

Continued

No.	Items	Quantities
23	Stainless steel thermometer dial	10
24	Pressure valve humidifier for oxygen cylinder	50
25	Finger clip oximeter	50
26	CPR compression plate	10
27	Flatcar for transshipment	10
28	Electronic laryngoscope	1
29	B-mode ultrasound scanner	1
30	Portable CT	1
31	Portable DR	1
32	Automated clinical chemistry analyzer (dry-type)	1
33	Specimen transport box	10

2. 2. 4 Medical Consumables

Quantity Standard of Medical Supplies

(Based on 10-day usage of 1,000 patients)

No.	Items	Quantities
1	Oxygen cylinder	50
2	Ultraviolet cabinet sterilizer	10

Continued

No.	Items	Quantities
3	Air sterilizer	140
4	Hand sanitizer	1,000
5	Sterilization monitoring test pack (pack)	10
6	Rubbing alcohol with 75% ABV (500 mL)	400
7	0.5% iodophor (50 mL)	100
8	Chlorine effervescent tablets	100
9	Disposable oxygen mask	300
10	Infusion set (automatic liquid stop)	500
11	Disposable tourniquet	500
12	Cotton swab (pack)	500
13	Disposable treatment towel (non-sterile)	100
14	Sterile treatment towel	100
15	Sterile gauze (pack)	1,000
16	Aseptic cotton ball (pack)	1,000
17	1 mL syringe	100
18	2 mL syringe	100
19	5 mL syringe	100
20	20 mL syringe	100
21	Urinary catheter	50
22	Urethral catheterization kit	50

Continued

No.	Items	Quantities
23	Medical gauze 7 * 9—2 per pack	500
24	Medical gauze 7 * 9 (x-ray detectable)—5 per pack	100
25	Disposable oxygen pipe (with humidifying liquid) 100 mL	50
26	Disposable oxygen pipe (with humidifying liquid) 250 mL	50
27	Blood glucose test paper; Blood glucose meters of the same brand	500; 12
28	ECG monitoring electrode	200
29	Gastric tube	50
30	Infusion applicator	500
31	Disposable vacuum for blood collection	500
32	Disposable needle for venous blood collection	300
33	Mercury thermometer	300
34	Stethoscope	100
35	Oxygen bag	50
36	Portable infusion stand	100
37	Stainless steel treatment plate	50
38	Kidney basin	50

Continued

No.	Items	Quantities
39	Soaking bucket (for sterilization of treatment plate and kidney basin)	50
40	Wheelchair	10
41	Weighing scale	5
42	Infrared temperature gun	20
43	Height scale	5
44	Electronic sphygmomanometer	30
45	Negatoscope (Single Couplet LED)	3
46	Negatoscope (Double Couplet LED)	3

2.2.5 Medication

Quantity Standard of Medicines

(Based on 10-day usage of 1,000 patients)

No.	Medicines	Specifications	Quantities
1	Moxifloxacin Hydrochloride Tablets	0.4 g * 3 T	1,000
2	Amoxicillin and Clavulanate Potassium Tablets	0.487 g	200
3	Ceftriaxone Sodium for Injection	1.0 g	200

Continued

No.	Medicines	Specifications	Quantities
4	Furosemide Injection (Furosemide)	20 mg * 2 mL	50
5	Atropine Sulfate Injection	1 mL : 0. 5 mg	50
6	Nicothamide Injection (Clonamine)	375 mg : 1. 5 mL	50
7	Deacetylhairoside Injection (Cediland)	0. 4 mg : 2 mL	50
8	Nifedipine Controlled Release Tablets	30 mg * 7 T	250
9	Nitroglycerin Tablets	0. 5 mg * 100 T	50
10	Methoxamine Hydrochloride Injection	1 mL : 10 mg	50
11	Noradrenaline Bitartrate Injection	2 mg : 1 mL	50
12	M-hydroxyamine Bitartrate Injection (Alamin)	1 mL : 10 mg	50
13	Amiodarone Hydrochloride Tablets	0. 2 g	50
14	Dopamine Hydrochloride Injection	2 mL : 20 mg	50

Continued

No.	Medicines	Specifications	Quantities
15	Adrenaline Hydrochloride Injection	1 mg : 1 mL	50
16	Isoprenaline Hydrochloride Injection	1 mg : 2 mL	50
17	Metoclopramide Hydrochloride Injection (Metoclopramide Injection)	1 mL : 10 mg	50
18	Compound Amino Acid Injection (18AA-V)	250 mL : 8.06 g total amino acids + 12.5 g xylitol	100
19	Isatis Root Granules	10 g * 15	1,000
20	Dexamethasone Sodium Phosphate Injection	1 mL : 5 mg	100
21	Potassium Chloride Sustained Release Tablets	0.5 g * 24	150
22	Montmorillonite Powder	3 g * 12	150
23	Diclofenac Sodium Suppositories	50 mg * 10	100
24	Vitamin B_1 Tablets	10 mg * 100	500
25	Vitamin B_6 Tablets	10 mg * 100	500

Continued

No.	Medicines	Specifications	Quantities
26	Vitamin C Tablets	100 mg * 100	150
27	Protamine Biosynthesis Human Insulin Injection (Novolin 30R Penfill)	300 IU : 3 mL	50
28	Lopinavir/Ritonavir Tablets	200 mg/50 mg * 120	350
29	Chloroquine Phosphate Tablets	0.25 g * 100	200
30	Ribavirin Injection	0.1 g	100
31	Ribavirin Tablets	100 mg * 25	2,000
32	Oseltamivir Phosphate Capsules	75 mg * 10	500
33	Sodium Chloride Injection (0.9% NaCl)	500 mL : 4.5 g	900
34	Sodium Chloride Injection (0.9% NaCl) 10 mL	10 mL : 90 mg	500
35	Jinye Baidu Granules		500
36	Lanqin Oral Liquid	10 mL * 6	500
37	Tylenol	20 per box	100

Continued

No.	Medicines	Specifications	Quantities
38	Hydrotalcite Chewable Tablets	0. 5 g * 20	100
39	Domperidone Tablets	10 mg * 42	100
40	Fenbide Capsule		100
41	Loratadine Tablets	10 mg * 6	100
42	Berberine Tablets		100
43	Amlodipine Tablets	5 mg * 7	200
44	Metoprolol Sustained Release Tablets	47. 5 mg * 7	100
45	Clopidogrel Tablets	75 mg * 7	100
46	Aspirin Tablet	0. 1 g * 30	100
47	Atorvastatin	20 mg * 7	100
48	Metformin Tablets	0. 5 g * 20	200
49	Estazolam Tablets	1 mg * 150	10
50	Budesonide Suspension for Inhalation	1 mg : 2 mL (5 per bag)	50
51	Ipratropium Bromide Solution for Inhalation	2 mL : 250 μg * 10	50
52	Aminophylline Sustained-release Tablets	0. 1 g * 24	10

Continued

No.	Medicines	Specifications	Quantities
53	Cefaclor Capsules	0. 25 g * 6 * 2	50
54	Huoxiang Zhengqi Soft Capsule	0. 45 g/capsule * 24	3,000
55	Jinhua Qinggan Granules	5 g per bag	2,000
56	Jinlianhua Soft Capsule	0. 55 g * 24	2,000
57	Powerful Loquat Dew	250 mL	2,000
58	Lianhua Qingwen Capsule	0. 35 g * 24	2,000
59	Omeprazole Sodium for Injection	40 mg	200
60	Magnesium Isoglycate Injection	10 mL : 50 mg	100
61	Raceanisodamine Hydrochloride Injection	10 mg * 1 mL	100
62	Lidocaine Hydrochloride Injection	0. 1 g * 5 mL	250
63	Abidol Hydrochloride Tablets	0. 1 g * 12 T	1,200
64	Vitamin C Injection	2 mL : 0. 5 g	2,000
65	Levofloxacin Lactate and Sodium Chloride Injection	0. 2 g * 100 mL	100

Continued

No.	Medicines	Specifications	Quantities
66	Glucose Injection (50% concentrated)	20 mL : 10 g	500
67	Glucose Injection (5%)	500 mL : 25 g	500
68	Glucose Injection (5%)	250 mL : 12.5 g	500
69	Glucose Injection (5%)	100 mL : 5 g	500
70	Glucose Injection (10%)	250 mL : 25 g	500
71	Glucose and Sodium Chloride Injection	500 mL : 25 g (5%)	500
72	Sodium Chloride Injection (0.9% NaCl)	250 mL : 2.25 g	500
73	Sodium Chloride Injection (0.9% NaCl)	100 mL : 0.9 g	500
74	Lobeline	3 mg : 1 mL	250
75	Compound Glycyrrhizin Injection	20 mL	100
76	Furosemide Tablet	20 mg * 100	100
77	Zolpidem Tablet	10 mg * 20 T	50
78	Indometacin Suppositories	0.1 g * 10	100
79	Yinhuang Buccal Tablets	0.65 g * 24 T	200

Continued

No.	Medicines	Specifications	Quantities
80	Cetirizine Hydrochloride Tablets	10 mg * 6 T	50
81	Cefdinir	0. 1 g * 10 S	150
82	Seretide	250 μg : 50 μg * 60 D	100
83	Oral Rehydration Salts Ⅲ	5. 125 g * 6	300
84	Berberine	0. 1 g * 100 T	50
85	Licorice Mixture	100 mL	500
86	Compound Methoxyphenamine Capsules	60 S	100
87	Acetaminophen	0. 3 g * 24 T	100
88	Diazepam Injection	2 mL : 10 mg * 10	150
89	Sodium Valproate	0. 2 g * 100 T	100
90	Betaloc	25 mg * 20 T	50
91	Acarbose Tablet	50 mg * 30 T	100
92	Zheng Chang Sheng Capsule		100
93	Suppositories Glycerol		300

Continued

No.	Medicines	Specifications	Quantities
94	Shufeng Jiedu Capsule/Granule		2,000
95	Qingfei Paidu Decoction *		According to the actual demand
96	Decoction No. 1 *		According to the actual demand
97	Decoction No. 2 *		According to the actual demand
98	Decoction No. 3 *		According to the actual demand

* Traditional Chinese medicine decoction.

The Logistics Company Enters the Red Cross Prevention and Control Material Warehouse (Responsible for the Logistics Management of Materials)

3 Management of Patients, Staff and Medical Supplies

3.1 Management of Patients

This section introduces three crucial criteria for patient management in a makeshift hospital: admission, referral, and discharge, with reference to *The New Coronavirus Infected Pneumonia Diagnosis and Treatment Plan (The Seventh Edition, Tentative)* formulated by the Center for Epidemic Prevention and Control in Wuhan.

3.1.1 Admission Criteria

Patients admitted to makeshift hospitals should satisfy the following conditions:

—With confirmed 2019-nCoV;

—Clinically classified as mild by *The New Coronavirus Infected Pneumonia Diagnosis and Treatment Plan (The Seventh Edition, Tentative)* issued by the National Health Commission of China;

—Aged 18 – 65 and having self-care ability;

—No underlying diseases related to the respiratory system, cardio-cerebral vascular system, and no mental illness;

—Negative in the influenza virus nucleic acid test.

Traditional Chinese Medicine Plays an Important Role in the Prevention and Control of COVID-19

3.1.2 Transfer Criteria

Patients classified as severe or critical should

immediately be transferred to the designated hospital for treatment. [1]The specific standards were as follows.

(1) Clinical classification for severe patients

Adults met any of the following conditions:

—Shortness of breath, RR≥30 times/min;

—Resting-state blood oxygen saturation≤93%;

—Arterial partial pressure of oxygen (PaO_2) /oxygen concentration (FiO_2) ≥300 mmHg (1 mmHg = 0.133 kPa). Should be corrected for high altitudes (above 1,000 meters) according to the following formula: (PaO_2/FiO_2) × [Atmospheric Pressure (mmHg)/760].

—Pulmonary imaging showed that the lesions progressed more than 50% within 24 – 48 hours.

Children met any of the following conditions:

—Shortness of breath (< 2 months of age, RR ≥ 60 beats/min; 2 to 12 months of age, RR≥50 beats/min; 1 to 5 years old, RR ≥ 40 beats/min; > 5 years old, RR ≥ 30 times/min) except for in fever or crying;

—Resting-state blood oxygen saturation≤92%;

—Signs of severe breathing difficulty (gasping, nasal flaring, rib retractions), cyanosis, and intermittent apnea;

---

① In addition, patients with comorbidities or concomitant diseases that required transfer to general hospitals should be transferred to designated general hospitals.

—Drowsiness and convulsions;

—Refusing to eat and having symptoms of dehydration.

(2) Clinical classification for critical patients

Patients met any of the following conditions:

—Respiratory failure and requiring mechanical ventilation;

—Shock;

—Combined organ failure and requiring ICU monitoring and treatment.

3.1.3 Discharge Criteria

Patients discharged should satisfy the following conditions[1]:

—The body temperature returned to normal for 3 days in a row;

—Significant improvement in respiratory symptoms;

—Pulmonary imaging showed significant improvement in acute exudation;

—Negative for two consecutive times in the nucleic acid test of sputum, nasal polyps, and other respiratory

---

[1] Discharged patients should follow *The New Coronavirus Infected Pneumonia Diagnosis and Treatment Plan (The Seventh Edition, Tentative)* to complete isolation and follow-up visits if necessary.

specimens (sampling interval≥24 hours).

3.2 Staff Training

The training included orientations that familiarized the staff with work in the makeshift hospitals and infection prevention and control. Particularly, instructions were given regarding the staff's health status.

—Frontline workers, who worked long hours wearing impermeable protective clothing, should be aware of cardiac arrest caused by hypokalemia and hypoglycemia. Drink potassium-rich beverages such as juices if necessary.

—Medical staff with cerebrovascular diseases should actively monitor blood pressure and take calcium antagonists, β-antagonists and other drugs when necessary, because high tension may have a sudden rise in blood pressure and cause severe symptoms such as cerebrovascular accident.

—Medical staff with sleep-disordered breathing should wear a CPAP ventilator during sleeping to prevent critical complications caused by severe hypoxia at night due to excessive fatigue.

In addition, training on emotional management might be provided to ensure the mental health of the staff. When necessary, assess the psychological status of medical staff, and conduct psychological counseling to medical staff in need.

Pharmaceutical Companies Providing Assistance to Fight Against COVID-19

3.3 Management of Medical Supplies

3.3.1 Allocation Rules and Regulations

The allocation standards of medical supplies in the makeshift hospitals:

—Should refer to the allocation standards of the regular infectious ward and intensive care unit.

—Should not be changed at will. Temporary adjustment required approval by clinical experts and leaders.

The demand for medical supplies should be reported daily.

3. 3. 2 Procurement Rules and Regulations

The procurement catalog, which was updated promptly according to the specific situation, should be strictly followed.

Detailed information on product model and specifications of medical supplies (especially protective supplies) should be disclosed to qualified suppliers.

The purchase quantity should meet the 10-day standard of the makeshift hospitals (bed occupancy rate = 100%).

3. 3. 3 Acceptance, Inspections and Recording Rules and Regulations

The review team was responsible for the acceptance and the inspection of the medical supplies to ensure that the products in the makeshift hospital are qualified.

During the inspection process, the review team should check and record the product name, supply unit, specifications, quantity, price, manufacturer, batch number, validity period, packaging, approval document number, etc. Particular attention should be paid to the validity of the invoices and the quality certification documents.

Pharmaceutical Companies Donated Drugs to Help Fight Against COVID-19

Acknowledgements

The study was funded by the National Natural Science Foundation of China (71704024, 81801778), National Key R&D Program of China (2018YFC0910700), and the Fundamental Research Funds for the Central Universities (CXTD11-01). We were also to all experts of the Center for Epidemic Prevention and Control in Wuhan for data sharing and consulting services.

References

[1] *The Law of the People's Republic of China on Prevention and Control of Infectious Diseases.*

[2] *The Hubei Provincial Makeshift Hospital Rescue Work Management Regulations.*

[3] *The Notice on Adjusting the Standards of Medical Supplies for the Opening of Makeshift Hospitals.*

[4] *The New Coronavirus Infected Pneumonia Diagnosis and Treatment Plan (The Seventh Edition, Tentative).*

[5] *The Makeshift Hospital In-Hospital Diagnosis and Treatment Plan (The First Edition).*

[6] *The National Essential Medicines List.*

[7] *The National Formulary of China.*

[8] *The Makeshift Hospital Management Rules.*

鸣　　谢

Acknowledgements

国药控股湖北有限公司

Sinopharm Group Hubei Co. , Ltd.

江苏恒瑞医药股份有限公司

Jiangsu Hengrui Pharmaceuticals Co. , Ltd.

上药科园信海医药湖北有限公司

SPH Keyuan Xinhai Pharmaceutical Hubei Co. , Ltd.

南京医药湖北有限公司

Nanjing Pharmaceutical Hubei Co. , Ltd.

九州通医药集团股份有限公司

Jointown Pharmaceutical Group Co. , Ltd.

(hereafter referred to as JOINTOWN)

后　记

武汉方舱医院从确保“应收尽收”，到“关门大吉”，托起了生命的方舟。武汉方舱医院的建设和运营，积累了应对重大突发公共卫生事件的丰富经验，希望本书能够成为总结经验的一个载体。

面对百年来全球发生的最严重传染病大流行，新中国成立以来我国遭遇的传播速度最快、感染范围最广、防控难度最大的重大突发公共卫生事件，方舱医院同时间赛跑、与病魔较量，迅速初步遏制疫情蔓延势头，为取得武汉保卫战、湖北保卫战的决定性成果发挥了重要作用。

武汉方舱医院的建设运营，是政府主导、社会协同的成功范例。本书的编写人员，全部是武汉方舱医院建设运营的参与者、见证者和研究者。由于参与编写的人数较多、语言风格不完全相同，本书的中英文版本，客观上存在少数语句意译或直译、表述不完全一致的情况，特此说明。

本书在编写过程中得到武汉大学、华中科技大学、武汉市卫健委、武汉市医保局相关专家和领导、部分医药企业的大力支持，在此一并表示感谢。

Postscript

The Wuhan makeshift hospitals have supported the ark of life from the policy "ensuring all receivables" to the closing. Related staff and organizations have accumulated rich experience in responding to major public health emergencies from the construction and operation of Wuhan makeshift hospitals. We wish this book can become a carrier for summing up experience.

China has encountered major public health emergencies with the fastest spreading speed, the widest range of infections, and the most difficult prevention and control since the foundation. Facing the most serious infectious disease pandemic in the world in a century, the makeshift hospitals raced against time, competed with the disease, quickly and initially curbed the spread of the epidemic, and played an important role in achieving decisive results in the defense of Wuhan and Hubei.

The construction and operation of Wuhan makeshift hospitals is a successful example of government-led collaboration with the social. The authors of this book are all participants, witnesses and researchers in the construction and operation of the makeshift hospitals. Due to the large number of people who participated in the compilation, the language styles are different, and objectively there are a few cases in which the free translation or literal translation and expression of the Chinese and English versions of this book are not completely consistent, and here is to explain.

This book has received strong support from relevant experts and leaders of Wuhan University, Huazhong University of Science and Technology, Wuhan Municipal Health Commission, Wuhan Healthcare Security Administration, and some pharmaceutical companies in the process of writing. The specific list will not be listed here, but thank you all.